Water Vapor, Not Carbon Dioxide, is Major Contributor to the Earth's Greenhouse Effect

Water Vapor, Not Carbon Dioxide, is Major Contributor to the Earth's Greenhouse Effect

Putting the KIBOSH on Global Warming Alarmists

Roy Cataldo

Copyright © 2014 by Roy Cataldo.

Library of Congress Control Number: 2014914658
ISBN: Hardcover   978-1-4990-6316-5
           Softcover   978-1-4990-6317-2
           eBook   978-1-4990-6318-9

All rights reserved. No part of this book may be reproduced or transmitted in any form or by any means, electronic or mechanical, including photocopying, recording, or by any information storage and retrieval system, without permission in writing from the copyright owner.

Any people depicted in stock imagery provided by Thinkstock are models, and such images are being used for illustrative purposes only. Certain stock imagery © Thinkstock.

This book was printed in the United States of America.

Rev. date: 10/16/2014

To order additional copies of this book, contact:
Xlibris
1-888-795-4274
www.Xlibris.com
Orders@Xlibris.com
669319

LIST OF TABLES

Table 1. Concentrations of selected greenhouse gases in parts per billion. Water vapor excluded. 27

Table 2. Concentrations (PPB) of man-made additions, total concentrations, and percentages. Water vapor excluded. 28

Table 3. Greenhouse gases, water vapor excluded, adjusted for heat retention relative to CO_2. 30

Table 4. Man-made and percentages of greenhouse gases, water vapor excluded, adjusted for heat retention relative to CO_2. 30

Table 5. Contributions to the "Greenhouse Effect" as a percent of total with and without water vapor. 34

TABLE OF CONTENTS

Preface 9

I. Defining why carbon dioxide is not the major contributor to the Earth's Greenhouse Effect. 9

II. A definition of the Greenhouse Effect. 11

III. Why only certain gases such as water vapor, carbon dioxide, etc., retain heat 12

IV. The reason water vapor overwhelms carbon dioxide in greenhouse effect. 12

V. Why global warming alarmists have difficulty accepting water vapor as an air "pollutant" or greenhouse gas. 14

VI. Which greenhouse gases are significant and are used in the following climate change studies. 15

Introduction 17
   Characteristics of Water Vapor 18
   Important greenhouse
   gas properties of water vapor 19
   A listing of greenhouse gases and their
   basic data sets to be used in the presented
   analysis of greenhouse effect importance. 21

Chapter I 25
   Water vapor not carbon dioxide
   is the cause of global warming. 25
   Table 1 27
   Table 2 28
   Table 3 30
   Table 4 30

Chapter II 32
   Water vapor – the most significant
   greenhouse gas. 32

Table 5 34

Conclusion 37
    The closed discussion of Global Warming 39
    Climate Myths 41
    Glacial Melting 42
    How global warming myths are created 43
    Global warming myth linking non-existing "torrential" rains and flooding 44
    Why man-made greenhouse gas additions are extremely small relative to the Pre-Industrial baseline 45

Epilogue 47

References 53

PREFACE

I. *Defining why carbon dioxide is not the major contributor to the Earth's Greenhouse Effect.*

In spite of the fact computer models used to support emission reductions required of the original Rio Treaty of 1992 (1), and the later Kyoto Protocol (2), resulted in warming increases which greatly exaggerated actual global warming, alarmists of today still believe the false notion that the carbon dioxide contribution to global warming is about 80% of all greenhouse gases considered. Using Department of Energy (DOE) (3a), Carbon Dioxide Information Analysis Center (CDIAC) (3b), Environmental Protection Agency

(EPA) (4), and European research data, it is provable that such a statement is false.

It will be shown the contribution of carbon dioxide is 72.369% only under studies where water vapor (H_2O) content of the atmosphere is set equal to zero. When studies are carried out with water vapor considered an air pollutant with its correct percentage content of the atmospheric gases set to 95.000% (5a) (5b), the actual carbon dioxide (CO_2) importance is reduced to a relatively small 3.618%.

Another significant finding of greenhouse gas relative importance in studies without and with water vapor included in the studies shows the total man-made greenhouse effect is reduced from 5.53% to 0.28% when studies without water vapor and with water vapor are conducted. In either case man-made greenhouse effect is relatively small and does not justify the carbon tax being proposed by global warming alarmists.

II. *A definition of the Greenhouse Effect.*

Most of the sun's energy directed at the earth is in the form of *short* wavelength radiation. This short wavelength radiation does not cause the greenhouse effect since the original energy from the sun is absorbed at or near the earth's surface, and is reradiated back to the atmosphere in the form of long wavelength or infrared radiation. Thermal energy or heat comes about by absorption of the long wavelength radiation by certain gases such as water vapor, carbon dioxide, methane, nitrous oxide, and CFC's and miscellaneous gases. The "Greenhouse Effect" is the retention of the long wavelength energy in the gases caused by the ability of warming gases such as water vapor and carbon dioxide to support greater concentrations. A simple example of this is the common knowledge that warm air holds more water vapor than cold air.

III. *Why only certain gases such as water vapor, carbon dioxide, etc., retain heat*

As previously stated, it is the long wavelength radiation reflected from the surface of the earth that causes certain greenhouse gases to heat up or retain heat.

In order for the greenhouse gases to absorb the incoming long wavelength radiation, their molecules must be of a certain structure and size to cause the molecules to execute resonant vibration. It is the vibration at resonance that causes the molecules of the specific gases to heat up.

IV. *The reason water vapor overwhelms carbon dioxide in greenhouse effect.*

Actual satellite temperature data taken over the past decades has shown a warming signal in the coldest and driest masses, which can affect humans. Mainly, they are the great high pressure systems that form in northwestern North America and also in Siberia in the winter season. Greenhouse warming is mainly warming of the coldest air masses known. Further, the satellite records show a slight cooling trend for the decades of available satellite records. None of the global temperature records that scientists use show any warming over the time span of 1990 to the year 2000 (6).

Over 95 percent of the greenhouse effect is due to water vapor and only 3.618 percent is due to carbon dioxide. Since both water vapor and carbon dioxide absorb the same form of reflected energy from the earth's surface, it doesn't matter how much carbon dioxide is in the atmosphere. In the case of the wettest, warmest air masses (warm air holds much more water vapor than cold air) since the air will contain higher concentrations of water vapor, which will absorb much of the reflected energy.

Introducing carbon dioxide into the coldest air masses, which are the driest air masses containing little water vapor, is the same as putting in water since warming occurs whichever gas is present, raising temperature and water vapor concentration. These facts are corroborated by satellite temperature data over the past years showing no global warming to exist.

V. *Why global warming alarmists have difficulty accepting water vapor as an air "pollutant" or greenhouse gas.*

Water vapor is an extreme problem for global warming alarmists intent on blaming the hydrocarbon fossil fuels, such as coal, oil, and natural gas, for so-called "Global Warming" because of the sheer magnitude of 95.000% importance, it exhibits relative to the other four important greenhouse gases. The second important characteristic providing difficulty is that it is mostly of natural sources since the earth's surface is composed of mostly water at 71.00% coverage and the man-made contribution is an insignificant 0.001%. This makes it difficult to impose a man-made global warming tax on the world's energy supply as has been done by claiming CO_2 to be the largest contributor to global warming.

It may be noted that as a result of the inaccuracies of the original computer models used to promote emission reduction at the time of the Rio Treaty and Kyoto protocol, the discourse in the warming debate was changed from "Global Warming" to "climate change," which is used to this date. The change was made by warming alarmists to maintain some semblance of credibility in the climate debate.

VI. *Which greenhouse gases are significant and are used in the following climate change studies.*

It is well known there are other forms of climate change cycles known as primary causes of climate cycles such as the sun's energy output and eccentricities in the earth's orbit, which are not the focus of the following studies. For our purposes we are considering what are termed the secondary effect of a warming atmosphere to support higher concentrations of gases. To this purpose we will consider the most important gases in the order of their importance to be water vapor, carbon dioxide, nitrous oxide, methane, and chlorofluorocarbons, plus miscellaneous gases (CFC's and miscellaneous).

INTRODUCTION

It may be noted the "Greenhouse Effect" is a secondary climate change effect associated with the retention of heat reradiated from the earth to the greenhouse gases. Further, the Department of Energy (DOE) as well as the Environmental Protection Agency (EPA) and European research groups have extensive data which can be used to prove, at least within the validity of the data, which of the greenhouse gases is of most significance in effecting climate change.

It can be noted that most of the scientific community consider the five significant greenhouse gases (7) to be water vapor, carbon dioxide (CO_2), methane (CH_4), nitrous oxide (N_2O), and CFC's and miscellaneous gases.

Characteristics of Water Vapor

Water is essential for life. It also plays an important role in the earth's thermal balance. The specific heat of water (the heat required to raise 1 gram of water by 1 degree centigrade) is higher than almost any other material. Also, there is a tremendous amount of heat released to the atmosphere when high temperature water vapor condenses to form liquid water. Water has a massive ability to absorb heat. Most people have experienced the moderating effect of the ocean and large lakes on the climate of nearby cities. Such masses of water are slow to cool in cold weather and slow to warm in hot weather. When fossil fuel is burned, water vapor (steam) as well as carbon dioxide is formed. On cold days the vapor is visible as it cools and condenses as it leaves the exhaust. There is tremendous capacity of water vapor (as steam) to store or release thermal energy, this principle being utilized by all fuel-based power generating plants, including nuclear.

Despite the important thermal properties of water vapor, the DOE did not include the pertinent data for water vapor in its greenhouse gas data sets (8).

Important greenhouse gas properties of water vapor

There are two very important data points for water vapor in the evaluation studies which establish the relative importance ranking for each of the five greenhouse gases considered. The first is that 95.000% of the greenhouse effect of the earth's atmosphere is attributable to water vapor.

This number is so large as to be devastating to "Global Warming" alarmists, who maintain carbon dioxide to be the most significant greenhouse gas. It should be obvious that if the 95.000% water vapor figure is correct, then the remaining four greenhouse gases, including carbon dioxide, can only exhibit 5.000% of the total greenhouse effect. This total undercuts claims by the entire "global warming" community that carbon dioxide contributes 80% of the total atmospheric greenhouse effect.

It can be noted leading geochemists at leading observatories have suggested water vapor is so important that much of the climate change of the last 10,000 (9) years may be due to water vapor.

The second important data point associated with the natural versus man-made characteristics of water vapor is its minimal man-made contribution to global warming. It should not be surprising that with 71% of the earth's surface being covered with water, 96.5% of which is oceans, water vapor is an astounding 99.999% of natural origin. This places an additional difficulty on CO_2 alarmists interested in taxing man-made global warming since if water vapor is the most significant greenhouse gas and it is 99.999% from natural causes, it is only possible for water vapor to be man-made to an insignificant magnitude of 0.001%.

It would be difficult to convince the world's public to support a man-made water vapor tax if they knew its greenhouse effect was only 0.001%.

A listing of greenhouse gases and their basic data sets to be used in the presented analysis of greenhouse effect importance.

    A.   Five greenhouse gases under study
        1.   Water vapor (H_2O)
        2.   Carbon dioxide (CO_2)
        3.   Methane (CH_4)
        4.   Nitrous oxide (N_2O)
        5.   Chlorofluorocarbons (CPC's) and miscellaneous gases
    B.   Concentrations of the greenhouse gases
        1.   Units used will be parts per billion (PPB)

C. Pre-industrial baseline – used to establish time period from which to measure natural additions and man-made additions – concentration changes from the baseline.

D. Global Warming Potential (GWP) – used to adjust for the heat trapping or heat retention of the various greenhouse gases relative to carbon dioxide (CO_2). New values of concentrations relative to CO_2 are obtained by multiplying the original concentration number by the GWP for the gas under consideration.

    1. GWP Values

        $CO_2 = 1$

        $CH_4 = 21$

        $N_2O = 310$

        CFC's and misc. gases = 25

II. Tables of greenhouse climate change data and results

    Table 1 – Greenhouse gas baseline concentration and natural additions from the baseline without water vapor

Table 2 – Man-made additions data and total concentration, percent of total, and percent of total man-made calculations, without water vapor

Table 3 – Pre-industrial base concentrations, and natural additions concentrations corrected to CO_2 levels by multiplying by appropriate GWP, without water vapor

Table 4 – Man-made concentration data corrected relative to CO_2 by GWP, total data, percent of total (corrected), and percent of total man-made corrected, all without water vapor

Table 5 – Comparison of greenhouse gas climate change importance with and without water vapor. Comparison of natural and man-made components with water vapor.

CHAPTER I

Water vapor not carbon dioxide is the cause of global warming.

Temperature records used by climatologists show that over the past 1,000 years there have been periodic warm and cold periods. The relatively recent warming trend since about 1850 was nothing more than a continuation of the warming following the Little Ice Age (10), which was centered about 1650, rather than a sudden increase after a period of relatively uniform temperatures. The temperature record since 1850 shows a temperature decline between 1940 and 1970, which undercuts the global alarmists' explanation that continuous exponential increase (the famous "hockey stick curve") in temperature and carbon dioxide prove global warming is due to carbon dioxide (11). The fallacy of that assertion is proven by the fact that measured carbon dioxide concentrations in ice cores indicate carbon dioxide concentration changes after temperature changes, not before, therefore carbon dioxide is the result, not the cause, of global warming.

The purpose of the following study is to determine with the aid of the most reliable data regarding temperature change presently available which of the five significant greenhouse gases has the most influence on climate change.

The study begins with the introduction of the concentration data for the greenhouse gases studied without including water vapor data. Note the Department of Energy (DOE) did not include water vapor data with that of the other gases in Table 1. The table also establishes baseline data from which added "pollutants" of both natural causes and man-made causes can be measured. This is extremely important to be able to accurately compute man-made contribution to climate change.

Table 1

Concentrations of selected greenhouse gases in parts per billion. Water vapor excluded.

Department of Energy Data

	Pre-Industrial Baseline	Natural Additions
Carbon Dioxide (CO_2)	288,000	68,520
Methane (CH_4)	848	577
Nitrous Oxide (N_2O)	285	12
CFC's & misc. gases	25	0
TOTAL	289,158	69,109

Table 2

Concentrations (PPB) of man-made additions, total concentrations, and percentages. Water vapor excluded.

	Man-made Additions	Total Concentration (PPB)	Percent of Total	Percent of Total Man-made
Carbon Dioxide (CO_2)	11,880	368,400	99.438%	3.207%
Methane (CH_4)	320	1,745	0.471%	
Nitrous Oxide (N_2O)	15	312	0.084%	
CFC's & misc. gases	2	27	0.007%	
TOTAL	12,217	370,484	100.000%	3.298%

The total concentration column is calculated simply by adding the pre-industrial baseline, natural additions, and man-made additions data for each gas from Table 1 and Table 2. It may be noted that for carbon dioxide, with water vapor set to zero, the importance is magnified to 99.438% of the total greenhouse effect. This is an erroneous conclusion since the calculations are not complete until water vapor is included.

One additional step is required before water vapor is introduced into the calculations. The additional step requires the introduction of what has been termed the Global Warming Potential (GWP). The reason for the GWP is the various gases do not have the same heat retention properties as carbon dioxide and therefore must be corrected for heat retention relative to carbon dioxide. This is accomplished by multiplying the data of Table 1 and Table 2 by the appropriate GWP value for each gas.

The Global Warming Potential (GWP) for each gas, from various researchers, are for CO_2, CH_4, N_2O, and CFC's, respectively 1, 21, 310, and 25. Carrying out the multiplication of Table 1 and Table 2 data by the appropriate GWP data results in concentration data corrected for heat retention relative to CO_2. The results are given in Table 3 and Table 4.

Table 3

Greenhouse gases, water vapor excluded, adjusted for heat retention relative to CO2.

Units (PPB)	Pre-Industrial Baseline Adjusted	Natural Additions Adjusted
Carbon Dioxide (CO_2)	288,000	68,520
Methane (CH_4)	17,808	12,117
Nitrous Oxide (N_2O)	88,350	3,599
CFC's & misc. gases	2,500	0
TOTAL	396,658	84,236

Table 4

Man-made and percentages of greenhouse gases, water vapor excluded, adjusted for heat retention relative to CO2.

Non-Dimential Units	Man-made Additions (Adjusted)	Total Relative Effect	Percent of Total (Adjusted)	Percent of Total Man-made (Adjusted)
Carbon Dioxide (CO_2)	11,880	368,400	72.369%	2.33%
Methane (CH_4)	6,720	36,645	7.199%	
Nitrous Oxide (N_2O)	4,771	96,720	19.000%	
CFC's & misc. gases	4,791	7,291	1.432%	
TOTAL	28,162	509,056	100.000%	5.53%

It should be noted without water vapor and correcting for heat retention relative to carbon dioxide drops the total percentage importance for carbon dioxide from 99.438% of Table 2 to 72.369% of Table 4 or to a level closer to the 80% claimed by some warming alarmists.

CHAPTER II

Water vapor – the most significant greenhouse gas.

Before presenting the data for climate change statistics with water vapor, it should be pointed out that the "Global Warming" debate is replete (12) with data similar to data developed in Chapter I without water vapor. This is the only type of data presented by warming alarmists and their supporting media. Essentially the same type of data is supported by experimental measurement of man-made "pollution" mass, measured from man-made sources such as power generation, transportation, manufacturing, and farming. Since the same greenhouse gases without water vapor are measured, exaggerated levels of importance for carbon dioxide are arrived at. As a consequence of this fact, the public has always been exposed to incorrect results relative to the significance of the gases measured. For the world's public to be apprised to the complete set of "Greenhouse Gas" statistics, the study carried out in Table 5 is presented.

Although it is not recognized by the media or the global warming alarmists, most climatologists (13) agree water vapor is responsible for 95 percent or greater of the earth's greenhouse effect. (See item III of the Preface.) Further, it is 99.999% of natural causes. Using 95.000% for percent of total adjusted for water vapor, the other four greenhouse gases can only contain 5.000% of the total, and percent man-made data for water vapor is 0.001%. Thus the previous data of Table 3 and Table 4 can be constructed into Table 5, which indicates a direct comparison of percentage of relative greenhouse effect without (setting water vapor equal to 0.000%) and with (setting water vapor equal to 95.000%).

Table 5

Contributions to the "Greenhouse Effect" as a percent of total with and without water vapor.

% Natural with Water Vapor	% Man-made with Water Vapor	Based on Concentrations (PPB) Adjusted for Heat Retention Properties	Percent of Total Without Water Vapor	Percent of Total with Water Vapor
94.999%	0.001%	Water Vapor (H_2O)	0.000%	95.000%
3.502%	0.117%	Carbon Dioxide (CO_2)	72.369%	3.618%
0.294%	0.066%	Methane (CH_4)	7.100%	0.360%
0.903%	0.047%	Nitrous Oxide (N_2O)	19.000%	0.950%
0.025%	0.047%	CFC's and misc. gases	1.432%	0.072%
99.72%	0.28%	TOTAL	100.000%	100.000%

The percentage man-made carbon dioxide with water vapor included in Table 5 is calculated by multiplying the percent of total with water vapor (3.618%) by the percentage of CO_2 concentration from Table 2 (11,880/368,400) = .03225. Thus, (.03618 x .03225) = 0.117%. This is the insignificant magnitude the Kyoto Protocol would further reduce by another 30% to 0.035%. It would thus penalize the world's fossil fuel energy system by imposing a carbon tax, which is eventually borne by the people, without affecting any climate change.

Table 5 shows that adding all the man-made sources with water vapor included amounts to an insignificant 0.28% of the Greenhouse Effect. Again, man-made CO_2 contribution is a mere 0.117%, undercutting the global warming alarmists. This is why alarmists will say water vapor is not included in warming studies. Their "reason" is the unscientific argument that "it's always been done that way." Table 5 proves conclusively, using DOE, EPA, European research data, water vapor and not carbon dioxide (CO_2) is the predominant greenhouse gas.

Again, if the further reduction of 30% for CO_2 reduction required by the Kyoto Protocol is imposed on the world's fossil fuel system, the final carbon dioxide level would be reduced to 0.035% importance and would have virtually no effect on atmospheric temperature change.

It's further worth repeating that if water vapor is not considered a "Greenhouse Gas" which is preordaining the results of either analytical or experimental greenhouse effect studies, the warming of carbon dioxide is magnified from its real world effect of 3.618% to an importance of 72.369%, which is near the 80% importance quoted by some alarmists.

Further, when total versus man-made data are computed with water vapor included, the actual carbon dioxide global warming effect is a mere 0.117%, which would be further reduced to the insignificant level of 0.035% by the reduction required by the Kyoto Protocol.

The erroneous calculation of warming data by ignoring water vapor allows justification for blaming the burning of fossil fuels for atmospheric temperature change and provides justification for imposing a carbon tax on the world's energy supply.

The tax would have no effect on the earth's temperature change and its tremendous cost would reduce the standard of living of the world's population since they would eventually end up paying the tax.

CONCLUSION

The early days of the global warming debate used incomplete computer models tied mostly to the burning of fossil fuels such as coal, oil, and natural gas. Water vapor was excluded though most climatologists have stated over 95 percent of the earth's natural greenhouse effect is from water vapor.

Its omission greatly exaggerated predictions of the earth's temperature rise though these results were the basis of the original Rio Treaty on global warming and also for the later Kyoto Protocol, both of which demanded significant reductions in CO_2, CH_4, and other similar greenhouse gases. To this day, in spite of existing data to the contrary, claims are made by global warming alarmists that carbon dioxide is responsible for 80% of greenhouse gas effect. Similarly, measuring various greenhouse gas mass without including water vapor shows similar results.

This high level of importance for CO_2 cannot be true, however, since both water vapor and carbon dioxide absorb the same type of long wavelength energy being reflected by the earth's surface, so it doesn't matter how much carbon dioxide is in the atmosphere so long as the total concentration of water vapor is high enough to absorb most of the energy. This will always be the case since for the wettest air masses, which are the warmest, warm air holds more water molecules than cold air masses.

Similarly, for the coldest air masses, which are much dryer and hold very little water vapor, the effect of introducing carbon dioxide is the same as introducing water, since the absorption of any long wave radiation will result in increased warming and water vapor.

Thus, the correct values for water vapor greenhouse effect is 95.00%, and the correct value for carbon dioxide total (natural plus man-made) greenhouse effect is a mere 3.62%.

The erroneous conclusions further validate the fact that the proper magnitude of water vapor was not included in their calculations, models, or measurements.

The closed discussion of Global Warming

The findings of the tremendous influence of water vapor in the greenhouse gas debate totally undercut the global warming alarmists' edict to the effect the "global warming discussion is closed." This is a ridiculous claim since science discussion is never closed, as can be shown by examining the use of either direct current (DC) or alternating current (AC) to power the electrical grids which are still in use today. It could have been claimed as "settled science" after Thomas Edison invented the light bulb and attempted to provide early power grids based on direct current (DC). George Westinghouse subsequently came along and with the aid of the Tesla (AC) motor and generator inventions as well as French patents on electrical transformers set up the AC power grids in use today.

This should have settled the "settled science" fiasco once and for all, but the "Global Warming" alarmists sought to revive it to establish their claim the burning of fossil fuels is responsible for "global warming." It may be noted true "scientists" who are in fact geochemists at prestigious observatories (13) believe water vapor and its changes over time have been responsible for much of the global warming of the last 10,000 years because of increasing water content in the atmosphere.

Though the total (natural plus man-made) carbon dioxide (CO_2) contribution to greenhouse effect is reduced from 72.369% to 3.618% when water vapor is correctly taken into account, the all important man-made contribution is reduced to an insignificant 0.117% level. Further, if the further reduction of 30% required of the Kyoto Protocol were imposed, the CO_2 man-made contribution would be reduced to a statistically insignificant level of 0.035%.

Neither of these levels would effect climate change.

Climate Myths

It should be noted pre-selecting greenhouse gases to be studied or modeled while excluding others of predominant magnitude such as water vapor, amounts to pre-selecting or "cherry picking" data to be presented to the world's public. Further, the resulting incorrect data can be used to conflate various events, such as "glacial melting" to create climate myths.

For example, even though the previous discussion has proven the true value of man-made carbon dioxide (CO_2) is a mere 0.117%, this number has never been discussed or presented to the general public. The only figure presented to the public regarding carbon dioxide is 80% or higher, with the result it is easy to blame glacial melting on carbon dioxide.

Glacial Melting

In reality, research has shown that between the last ice age and the present interglacial period the atmospheric water vapor content has doubled, indicating water vapor, not carbon dioxide (as claimed by alarmists), is responsible for glacial melting.

Since water vapor is mostly of natural and not man-made origin, alarmists cannot blame water vapor is responsible since that would undercut their carbon tax argument.

Another example of the climate myth of man-made glacier melting was provided by Vice President Gore on a visit to Montana Glacier National Park in 1997 (14), the purpose of which was a photo op. In spite of the fact the park's own literature documented the glaciers had been melting throughout the past 150 years, Gore misled the public about natural warming during the 19th century as temperatures recovered from the frigid "Little Ice Age." Without any explanation of the natural causes of melting, he left the implication of man-made causes of melting by simply pointing to the glacier and stating to reporters, "This glacier is melting."

How global warming myths are created

This can be illustrated by examining how man-made global warming was used to imply warming makes hurricanes more frequent and stronger. In a global warming town meeting in Chapel Hill, North Carolina, members were told by Eileen Claussen (15) that Hurricane Fran was typical of what one could expect from global warming. The idea was further cemented in the general public's mind when insurance companies following the lead of Ms. Claussen used the town hall message to increase premiums. Their reason was increasing claims due to worsening hurricanes.

During the same time period a second panel, the IPCC (16) (17) (International Panel on Climate Change) had shown data that annual average winds in hurricanes occurring in the Atlantic Basin had statistically significant reductions over the past 50 years.

Further, other research over the same time period indicated the number of intense hurricanes declined in a statistically significant manner.

Global warming myth linking non-existing "torrential" rains and flooding

Another global warming falsehood was perpetrated, again by Vice President Gore, in a 1995 "Earth Day" speech (18). In the speech he declared, "Torrential rains have increased in the summer during agricultural growing seasons."

He was referring to as yet unpublished data, later published (19), claiming 2% increase for rain in the United States.

Two years later a Department of Commerce press release had magnified the original 2% increase to a whopping 20% increase. The reality was the amount of rain falling from the type of storms being discussed in actual rainfall measurements was only .6 inches.

Such a low level of rainfall increase does not cause flooding, as was being intimated by Gore's "Earth Day" "torrential rain" speech.

Why man-made greenhouse gas additions are extremely small relative to the Pre-Industrial baseline

As has been mentioned in the course of the previous studies, man-made additional greenhouse effect arises from such human activities as power generation, manufacturing, transportation, and farming. The reason such an impressive list of activities does not overwhelm the greenhouse effect is simply that the magnitude of the additional concentration of "pollutants" added to the atmosphere is relatively insignificant relative to the magnitude of "pollutants" that originate in the atmosphere at the time the original Pre-Industrial baseline concentrations were established. This can be verified by examining Table 1 and Table 2 of the previous study.

In other words, most of the "pollution" was already in the atmosphere established by nature before man started creating his own "pollution."

So why did nature dump so much "pollution" in the atmosphere as to be responsible for most of today's greenhouse effect? The answer is simply to provide a heat shield over the earth, since without it, the surface of the earth would be 86 degrees Fahrenheit colder, too cold for most life before or after the artificial Pre-Industrial baseline time setting.

EPILOGUE

I.   *Making sense of the front cover.*

The purpose of the front cover is to encapsulate the basic findings of the enclosed studies.

The lines emanating from the sun towards the earth's surface represent long wavelength solar radiation, which is absorbed by the earth, water vapor, and other gases. A small portion of the incoming short wavelength radiation is reflected by the earth's surface back into space.

Most of the incoming radiation is absorbed and converted into long wavelength infrared radiation near or on the earth's surface. Heat results from the ability of some atmospheric gases, such as water vapor, carbon dioxide, methane, nitrous oxide, and miscellaneous gases whose molecules can vibrate to the reflected long wave radiation and become heated. Greenhouse effect is simply the retention of the generated heat in atmospheric greenhouse gases.

The Greenhouse Effect is essentially nature's protective thermal blanket around the earth based upon baseline concentrations, since without it the earth's surface would be 86 degrees Fahrenheit colder.

II. *The meaning of the enclosed data within circles.*

The circles enclosing the cloud designations indicate clouds may both emit and reflect energy from the incoming sources.

The circles enclosing the water vapor and carbon dioxide designations indicate results of a comparison of these two important greenhouse gases. The percentage numbers within the circles indicate water vapor is by far the most powerful greenhouse gas in the atmosphere.

Reading from top to bottom the printed information indicates the percentages from the total greenhouse contribution to the man-made contribution for the two greenhouse gases.

The total greenhouse contribution is described within the body of this study. Briefly, it includes the pre-baseline contribution, plus the additional contribution due to natural causes, plus additional man-made causes.

It may be noted carbon reductions required of the climate change agreement known as the Kyoto Protocol are indicated for carbon dioxide. The Kyoto Protocol is an international agreement created under the United Nations Framework Convention on Climate Change (UNFCC) in Kyoto, Japan in 1997.

The conclusion from the comparison of water vapor versus carbon dioxide indicates water vapor, which has a total greenhouse contribution of 95.000% and man-made contribution of a mere 0.001%, is the most powerful greenhouse gas when compared to carbon dioxide and other greenhouse gases. Further, the insignificant 0.001% man-made contribution of water vapor indicates not much can be done to further reduce the man-made contribution of water vapor. Carbon dioxide, on the other hand, has a total greenhouse contribution of a mere 3.62%, indicating very little greenhouse effect. Carbon dioxide is also shown to have a small man-made greenhouse contribution of only 0.117%. It is important to know President Obama is presently calling for a 30% carbon cap reduction, which is equivalent to the original Kyoto Protocol reductions. It would do nothing for climate change since it would reduce the 0.117% man-made contribution of carbon dioxide to 0.035%, which is a number equivalent to that of statistical variability.

The Kyoto Protocol reduction comparison for carbon dioxide is shown in the bottom two circles of the carbon dioxide column on the front cover.

Thus, the information on the front cover comparing the important total and man-made contributions of water vapor versus carbon dioxide in the greenhouse gas debate indicate water vapor by far is the most powerful gas in the greenhouse gas system.

REFERENCES

(1) Rio Treaty, also known as the Earth Summit. See Earth Summit – Wikipedia, the free encyclopedia

(2) Kyoto Protocol – Wikipedia, the free encyclopedia.

(3a) Recent Greenhouse Gas Concentrations DOE/ER-236, U.S. Department of Energy, Washington, D.C.

(3b) Carbon Dioxide Information Analysis Center(CDIAC) Oak Ridge National Laboratory, Oak Ridge, TN 37831-6290 Recent Greenhouse Gas Concentrations, DOI: 10.3334/CDIAC/atg.032, updated February 2014.

(4) Global Warming Potential Greenhouse Gases Climate Change, U.S. EPA Atmospheric Greenhouse Gas

Concentrations (CSI013/Climo52 ... Greenhouse Gas – Wikipedia, the free encyclopedia.

(5a) Water Vapor 95% *The Physics of Atmospheres,* 2nd edition, Cambridge University Press, 2002.

(5b) Greenhouse Gases, Water Vapor and You – MIT Joint Program on the globalchange.mit.edu/mews/news-item.php?id-139 Fact that 95% of all 'greenhouse" gases is water vapor.

(6) 1990 IPCC predictions Confront the Data, Clive Best www.clivebest.com/?p=2208

(7) Greenhouse gas – Wikipedia, the free encyclopediaen.wikipedia.org/wiki/greenhouse_gas

(8) Greenhouse Gases Overview – Environmental protection Agency *www.epa.gov/climatechange/ghgemissions/gases.html*

(9) The Geologic Record and Climate Change – Pete's Place – Blogger petesplace-peter.blogspot.com/.../geologic-record-and-climate-change.html.

(10) Little Ice Age – Wikipedia, the free encyclopedia. en.wikipedia.org/Little_Ice_Age

(11) Water vapor seen as cause of rapid climate change *en.wikipedia/wiki/Hockey_stick_controversy*

(12) Climate Change, U.S. EPA – EPA *www.epa.gov/climatechange/* go to global greenhouse emissions data link

(13) Water vapor seen as cause of rapid climate change. *en.wikipedia.org/wiki/Water_vapor*

(14) Gore's defense of glacier tourism – junkscience.com junksciencearchive.com/news/sepp.html

(15) "World Climate Report" Moving on! *en.wikipedia.org/wiki/Eileen_Clauseen*

(16) Gross Errors in the IPCC-ARA Report concerning past and implied ... tropical.atmos.colostate.edu/includes/documents/.../gray2011.pdf.

(17) Tropical Cyclone Climatology *www.prh.noaa.gov/cphc/pages/FAQ/*

(18) Gore Shouts "Fire" in Crowded Greenhouse kevincraig.us/greenhouse.htm

(19) "Trends in high-frequency climate variability in the twentieth Century." *Nature,* 337 (1995), pp. 217-220.

www.ingramcontent.com/pod-product-compliance
Lightning Source LLC
Chambersburg PA
CBHW021042180526
45163CB00005B/2239